Abdelhafid Mimouni

Towards Sustainable and Innovative Chemistry

Abdelhafid Mimouni

Towards Sustainable and Innovative Chemistry

ScienciaScripts

Imprint

Any brand names and product names mentioned in this book are subject to trademark, brand or patent protection and are trademarks or registered trademarks of their respective holders. The use of brand names, product names, common names, trade names, product descriptions etc. even without a particular marking in this work is in no way to be construed to mean that such names may be regarded as unrestricted in respect of trademark and brand protection legislation and could thus be used by anyone.

Cover image: www.ingimage.com

This book is a translation from the original published under ISBN 978-620-6-72657-9.

Publisher:
Sciencia Scripts
is a trademark of
Dodo Books Indian Ocean Ltd. and OmniScriptum S.R.L publishing group

120 High Road, East Finchley, London, N2 9ED, United Kingdom
Str. Armeneasca 28/1, office 1, Chisinau MD-2012, Republic of Moldova, Europe
Printed at: see last page
ISBN: 978-620-8-26823-7

TOWARDS SUSTAINABLE AND INNOVATIVE CHEMISTRY

AUTHOR : DR. ABDELHAFID MIMOUNI

I am a freelance researcher in bioinorganic chemistry, specialising in macromolecular synthesis and characterisation. I hold a PhD from the University of Paris XII (1997) and a Diplôme des Études Approfondies en Systèmes Bioinorganiques from the University of Paris XI (1993), where I also obtained my Bachelor's and Master's degrees in Chemistry.

SUMMARY

This book explores sustainable chemistry and the role of inorganic materials in the transition to environmentally friendly energy solutions. It defines sustainable chemistry and its principles, traces the evolution of sustainable materials, and highlights the importance of metal complexes in catalysis. Water electrolysis technologies for hydrogen production are also examined, as are the challenges associated with their use, such as cost and toxicity. In addition, the book discusses the societal impact of sustainable technologies, highlighting the importance of social acceptability and sustainability policies. Finally, it proposes a way forward based on innovation, interdisciplinary collaboration and public awareness, to turn challenges into opportunities and foster a sustainable future.

TABLE OF CONTENTS

INTRODUCTION

The quest for a sustainable future has become one of the top priorities of our time. As environmental challenges intensify, it is imperative to explore innovative solutions that can reduce our ecological footprint while meeting our growing energy needs.

It is in this context that sustainable materials chemistry is emerging as a crucial field, combining fundamental science with practical applications. The aim of this book is to explore in depth the role of metal complexes and bio-based materials in the creation of sustainable solutions. We begin with catalysis, where transition metal complexes can improve the efficiency of chemical reactions, contributing to more environmentally friendly processes.

In particular, we will examine how these catalysts are essential in the conversion of water into hydrogen and oxygen, a key process for the future of renewable energies.

We will also address the issue of bio-based materials, highlighting innovations that make it possible to use renewable resources to develop sustainable composites. By incorporating metal complexes into these materials, we can not only improve their properties, but also open up new avenues for their use in a variety of applications. Finally, we will explore the challenges and prospects associated with the implementation of these new technologies. technologies.

Technical, economic and social obstacles must be overcome to ensure a transition to a sustainable energy future. We will highlight the importance of social acceptability and sustainability policies that support these innovations. Through this book, we hope to offer a clear and inspiring vision of current advances and future prospects in the field of sustainable materials chemistry. Whether you are a researcher, a student or simply passionate about environmental issues, we invite you to dive into this fascinating universe where science and sustainability meet.

CHAPTER 1

THE FOUNDATIONS OF SUSTAINABLE CHEMISTRY

Sustainable chemistry is a concept that forms part of a broader approach to sustainable development. It aims to design chemical processes and products that minimise environmental impact while meeting society's needs. In this chapter, we will define sustainable chemistry, trace its historical development and outline its fundamental principles.

1.1 Definition of sustainable chemistry

Sustainable chemistry is an emerging discipline at the intersection of chemical science and environmental imperatives. It is defined as a framework that integrates environmental, economic and social considerations at all stages of the design, manufacture and use of chemicals. This field of study aims to transform the way chemicals are developed and used, with a focus on sustainability and responsibility.

Objectives and principles

The fundamental aim of sustainable chemistry is to design chemical solutions that not only meet the needs of modern society, but do so in an efficient and environmentally friendly way. This involves a number of commitments:

1. **Reducing consumption of non-renewable resources**: Sustainable chemistry encourages the use of renewable raw materials, reducing dependence on fossil fuels. This includes initiatives to replace petroleum-based chemicals with bio-based alternatives, contributing to a circular economy.
2. **Reducing waste production**: One of the pillars of sustainable chemistry is waste prevention. This means that chemical processes must be designed to minimise by-products and optimise the use of reagents, leading t o increased efficiency and reduced waste.
3. **Limiting harmful effects on human health and the environment**: Sustainable chemistry aims to create chemical products that are less toxic and safer. This includes finding substitutes for hazardous substances, designing products that break down without harming the environment and assessing the environmental impact of products

throughout their life cycle.

Socio-economic importance

Sustainable chemistry is not limited to environmental concerns; it also encompasses economic and social dimensions. By promoting responsible practices, it can :

- **Stimulating innovation**: Research into sustainable chemistry encourages the development of new technologies and materials, creating economic opportunities.
- **Responding to societal expectations**: With growing awareness of environmental issues, consumers and companies are looking for products that reflect sustainable values. Sustainable chemistry meets this demand by incorporating sustainability criteria into the development of new products.

In short, sustainable chemistry represents an essential development in the chemical industry. By integrating environmental, economic and social considerations, it makes it possible to design solutions that are not only scientifically viable, but also respectful of our planet and its resources. Through this discipline, scientists have the opportunity to transform chemistry into a driver of positive change, contributing to a more sustainable future for generations to come.

1.2 History and development of sustainable materials

The idea of sustainability in the chemical industry took off in the 1990s, at a time when awareness of environmental issues was growing. This development was fuelled by a number of ecological crises, including increasing pollution, climate change and the depletion of non-renewable resources. Faced with these challenges, the search for innovative and sustainable solutions became increasingly important.

The 1990s: Awareness and Initiatives

In the 1990s, movements such as green chemistry began to emerge, encouraging scientists to design chemicals and processes that minimise environmental impact. The first conferences on green chemistry, such as the American Chemical Society conference in 1998, played a crucial role in spreading these ideas. These discussions highlighted the need to rethink chemical production methods and move towards more sustainable practices.

The 2000s: Technological progress and innovation

During the 2000s, research into sustainable materials gained momentum. Bioplastics, for example, began to appear on the market. Made from renewable raw materials such as corn starch or sugar cane, these plastics aim to reduce dependence on oil derivatives. Companies such as NatureWorks have pioneered the development of bioplastics, paving the way for a new era of environmentally-friendly materials. At the same time, the circular economy has gained momentum as a key concept, promoting reuse, recycling and waste reduction in all sectors, including chemicals. Initiatives have been put in place to integrate product life cycles that take into account their environmental impact at every stage, from design to disposal.

2010 to present: Towards widespread adoption

Over the past decade, pressure to adopt sustainable practices has led to significant advances in the field of sustainable materials. Many sectors, from packaging to construction, have begun to adopt innovative solutions, such as biomass composites. These materials, which combine renewable resources with inorganic components, offer improved performance while reducing environmental impact. Research into composite materials based on natural fibres, such as flax or jute, has also gained in importance. These light, strong materials are used in a variety of applications, including the automotive and aerospace industries, showing that durability can go hand in hand with performance. The development of sustainable materials reflects a paradigm shift in chemistry, from a focus on volume and cost to a more responsible and sustainable approach. This transformation is essential if we are to meet current and future environmental challenges. Progress in this area is a testament to the continued innovation and ability of the scientific community to meet the needs of society while preserving our planet.

1.3 Principles of green chemistry

Green chemistry is guided by twelve fundamental principles aimed at making chemical processes more environmentally friendly. These principles provide a framework for the research and development of sustainable chemical solutions. Here is a detailed development of some of the main principles:

1. Waste prevention

Waste prevention is the first principle of green chemistry. It involves designing chemical processes that minimise or eliminate waste production from the outset. This can involve :

- **Reaction optimisation**: Designing reactions that use reagents efficiently in order to reduce by-products.
- **Recycling technologies**: Incorporating methods for recovering and reusing materials to transform waste into resources.

2. Energy saving

Saving energy is essential to reducing the carbon footprint of chemical processes. This principle encourages :

- **Energy-efficient synthesis methods**: Use reactions that take place at room temperature or that require less energy to carry out.
- **Catalysis**: Using catalysts to increase the speed of reactions,

thereby reducing the time and energy required.

3. Use of renewable raw materials

Favouring renewable raw materials over non-renewable resources is a key aspect of green chemistry. This includes :

- **Biomass**: Using materials of biological origin, such as starch or cellulose, to produce plastics and other materials.
- **Sustainable alternatives**: Replacing petroleum-based chemicals with compounds derived from renewable resources, thereby contributing to long-term sustainability.

4. Reducing toxicity

Designing less harmful chemical products is a fundamental principle of green chemistry. This principle involves :

- **Risk assessment**: Analyse the toxicological properties of chemical products at the design stage to avoid those that present risks to human health or the environment.
- **Designing less dangerous chemicals**: Developing alternatives to

commonly used toxic substances, which may include research into less harmful solvents or harmless reagents.

So these principles of green chemistry are not just theoretical guidelines, but practical tools that can transform the way in which chemical products are designed and manufactured. By adopting these principles, chemists can contribute to a more sustainable future, reducing the environmental impact of chemical products while meeting the needs of society.

These principles serve as the basis for research and innovation in the field of sustainable chemistry, enabling the creation of solutions that respond to contemporary challenges.

CHAPTER 2

METAL COMPLEXES AND SUSTAINABLE CATALYSIS

Chapter 2 discusses the importance of metal complexes in sustainable catalysis, a key area for the development of more efficient and environmentally friendly chemical processes. By examining the principles of catalysis, the role of transition metals and concrete applications, this chapter highlights how these catalytic systems can contribute to sustainable chemistry.

2.1 Principles of catalysis

Catalysis is a phenomenon that speeds up a chemical reaction without the catalyst itself being consumed in the process. It can be classified into two main categories: heterogeneous catalysis and homogeneous catalysis.

- **Heterogeneous catalysis**: In this type of catalysis, the catalyst is in a different phase to the reactants. For example, in reactions where a solid catalyst is used for gases or liquids, the transfer of matter between the phases can be a limiting factor. Heterogeneous catalysis is often used in industrial processes because of the ease with which the catalyst can be recovered.

- **Homogeneous catalysis**: Here, the catalyst is in the same phase as the reactants, generally in the form of a solution. This approach offers advantages in terms of selectivity and efficiency, but separating the catalyst after the reaction can be more complex. Complexes play a crucial role in this type of catalysis, enabling reactions to take place at ambient temperature and under mild conditions.

2.2 Transition metals in catalysis

Because of their ability to adopt different oxidation states and form a variety of complexes, transition metals are essential in catalysis. They play many roles in chemical reactions:

- **Activation of reagents** : Metal complexes can activate reactive molecules by lowering the activation energy required for a reaction to take place. This results in an increase in the rate of reaction.
- **Selectivity control**: By modifying the structure of the metal complex, it is possible to direct the reaction towards specific products,

thereby minimising undesirable by-products.

- **Stability and recyclability**: Many metal laminates are designed to be stable and can be recycled after use, which is a major advantage for sustainable processes.

2.3 Practical applications

Metal complexes have many practical applications, particularly in the following areas:

- **Hydrogen production by electrolysis**: Catalysts based on metal complexes, such as nickel and platinum, are essential for improving the efficiency of electrolysis reactions. These catalysts facilitate the dissociation of water into hydrogen and oxygen, a key step in producing hydrogen as a clean energy carrier.
- **Reactions to convert CO2 into chemical products**: Metal complexes are also used to convert carbon dioxide into useful chemical products, such as methanol. This CO2 recovery process helps to reduce greenhouse gas emissions while producing renewable raw materials.
- **Case studies: Specific reactions using metal complexes**: Case studies illustrate the use of metal complexes in specific reactions, such

as the reduction of carbonyl compounds or polymerisation. These examples demonstrate the flexibility and efficiency of metal complexes in various chemical contexts.

In summary, metal complexes and their role in sustainable catalysis represent a promising avenue for modern chemistry. By integrating efficient and environmentally-friendly catalytic methods, it is possible to respond to today's energy and environmental challenges. This chapter has highlighted not only the principles and mechanisms of catalysis, but also concrete applications that demonstrate the importance of metal complexes in the transition to sustainable chemistry.

CHAPTER 3

BIOBASED MATERIALS

This chapter focuses on biobased materials, which represent a promising alternative to conventional materials derived from oil. By using renewable resources and minimising environmental impact, biobased materials play an essential role in the transition to a sustainable economy. We will examine their definition, the sources of biomass, and concrete applications illustrating their potential.

3.1 Definition and characteristics: What is a biobased material?

A biobased material is defined as any material whose components, either in part or in their entirety, are derived from renewable biological resources. These materials can be used in a variety of fields, from construction to packaging and textiles. The main characteristics of biobased materials include :

- **Renewability**: Unlike petroleum-based materials, biobased materials are derived from renewable resources, which means they can be produced sustainably.

- **Biodegradability**: Many biobased materials can decompose naturally, reducing waste and environmental impact.
- **Improved properties**: Thanks to technological advances, biobased materials can offer comparable or even superior performance to traditional materials.

3.2 Biomass sources: Types of biomass and their potential

Biomass sources are varied and include several types of organic matter:

- **Agricultural waste**: crop residues, straw and plant stems that can be transformed into bio-based materials.
- **Forestry waste**: Sawdust, shavings and other residues from the wood industry, which can be converted into composites or bioenergy.
- **Food waste**: leftovers from fruit, vegetables and other foods can be recycled to produce bioplastics or composites.
- **Energy crops**: Some plants, such as hemp and miscanthus, are grown specifically for their material properties.

These biomass sources offer huge potential for the production of biobased materials, helping to reduce waste and make the most of resources.

3.3 Practical applications

Biobased materials have concrete applications in a variety of sectors. Here are just a few examples:

- **Creating composites from organic waste**: Innovative techniques are being used to transform organic waste into composites, which can be used in a variety of applications, including such as furniture and construction panels. These composites not only reduce waste, but also offer sustainable alternatives to synthetic materials.
- **Case study : Biobased materials used in construction or packaging**: Several exemplary projects illustrate the use of biobased materials in construction, such as the use of particleboard made from wood residues or bioplastics for food packaging. These examples show how bio-based materials can be effectively integrated into traditional sectors, thereby reducing environmental impact.
- **Innovation in composite materials: Enhancing durability by integrating metal complexes**: Recent research is exploring the integration of metal complexes into bio-based matrices, thereby improving the strength and durability of composites. For example,

metal complexes can be added to reinforce mechanical properties or to impart specific functionalities, such as UV resistance or self-cleaning.

In short, bio-based materials represent an innovative solution to today's environmental challenges. By exploiting renewable resources and incorporating sustainable approaches, they offer viable alternatives to traditional materials. This chapter has highlighted not only the definition and characteristics of biobased materials, but also the role of biobased materials. They also demonstrate their potential in a range of applications, reinforcing their role in sustainable chemistry.

CHAPTER 4

RENEWABLE ENERGIES AND SUSTAINABLE MATERIALS

This chapter looks at the interconnection between renewable energy and sustainable materials, highlighting how these two fields can work together to support a sustainable energy transition. We will explore sustainable energy technologies, in particular solar cells, and the importance of metal complexes in their operation.

4.1 Sustainable energy technologies: Introducing renewable energies

Renewable energies are energy sources that regenerate naturally and are less likely to deplete the planet's resources. The main renewable energy technologies include :

- **Solar energy**: Conversion of sunlight into electricity using solar panels. This technology can be applied to photovoltaic or thermal solar cells.
- **Wind power**: The use of turbines to convert the kinetic energy of the wind into electricity. This type of energy production is growing rapidly and represents a significant proportion of the world's energy

mix.

- **Hydropower**: harnessing the energy of watercourses to produce electricity via dams or hydroelectric installations.

- **Biomass and biofuels**: Transformation of organic matter into energy, whether in the form of heat, electricity or fuels.

These technologies play a crucial role in reducing greenhouse gas emissions and combating climate change, while providing sustainable energy solutions.

4.2 Solar cells

Solar cells are one of the most promising applications of renewable energy, and the use of metal complexes is essential to improve their efficiency and durability.

- **Use of metal complexes in organic and inorganic solar cells**: Metal complexes are integrated into different types of solar cells, both organic and inorganic. In organic solar cells, these complexes can be used as electron acceptors, increasing the conversion of light into electricity. In inorganic solar cells, such as silicon-based cells, metal complexes can be used as catalysts to optimise the chemical reactions involved in energy conversion.

• **Case studies: Improved efficiency thanks to specific catalysts**

: A number of studies have shown that the integration of specific catalysts, such as those based on platinum or ruthenium complexes, can significantly improve the efficiency of solar cells. For example, research has shown that solar cells using metal complexes in their architecture can achieve higher energy conversion efficiencies than conventional cells. These advances illustrate the potential of metal complexes to revolutionise the field of solar technologies.

In short, renewable energies, particularly through solar cells, represent an essential avenue for a sustainable energy future. Integrating sustainable materials and metal complexes into these technologies not only improves energy efficiency, but also addresses environmental issues. This chapter has highlighted the importance of synergies between renewable energies and sustainable materials, paving the way for promising innovations in the global energy transition.

CHAPTER 5

CONVERSION OF WATER INTO HYDROGEN AND OXYGEN

This chapter explores the conversion of water into hydrogen and oxygen through electrolysis, a promising technology for the development of a sustainable energy economy. Focusing on the role of hydrogen as an energy carrier, we examine the principles of electrolysis, the catalysts required, the technologies employed and the potential applications of the hydrogen produced.

5.1 Introduction to Water Electrolysis

- **Definition and principle of electrolysis**: Water electrolysis is an electrochemical process that breaks down water into hydrogen and oxygen by applying an electric current. This is done in an electrolyser, where the water is subjected to a voltage that allows the water molecules (H2O) to dissociate. The hydrogen is collected at the cathode, while the oxygen is released at the anode.
- **Importance of hydrogen as a sustainable energy carrier**: Hydrogen is seen as a promising energy carrier because of its ability

to store and transport energy. It can be used in a variety of sectors, from power generation to mobility, and helps to reduce greenhouse gas emissions when produced from renewable sources.

5.2 Catalysts and Metal Complexes

- **5.2.1. Types of catalyst**: Catalysts play a crucial role in improving the efficiency of electrolysis. Precious metals, such as platinum and iridium, are often used because of their high catalytic activity. However, alternative metal complexes, such as nickel and cobalt, are gaining in popularity as more durable and economical options.
- **5.2.2 Recent innovations** : New strategies and materials are emerging to improve the efficiency of catalysts. These include the development of new catalytic composites, the optimisation of the structure of metals on a nanometric scale, and the use of innovative supports that increase the active surface area of catalysts.

5.3 Electrolysis Technologies

- **5.3.1. Alkaline electrolyser**: This type of electrolyser uses an alkaline solution as the electrolyte. It is relatively simple and economical, but less efficient than other technologies. Applications include large-scale industrial hydrogen production facilities.

- **5.3.2. Proton exchange membrane electrolysis (PEM)**: This technology uses a polymer membrane as the electrolyte, enabling high-efficiency electrolysis under the following operating conditions softer. The benefits include greater efficiency and faster response times, but the challenges include the high cost of materials.
- **5.3.3. High-temperature electrolysis**: This method uses electrolysers operating at high temperatures to improve energy efficiency. Future prospects include its use in conjunction with renewable heat sources, increasing the sustainability of the process.

5.4 Hydrogen Product Applications

- **Use in hydrogen fuel cells and hydrogen-powered vehicles**: Hydrogen produced by electrolysis can be used in hydrogen fuel cells, offering a clean alternative to combustion-powered vehicles. This reduces CO2 emissions in the transport sector.
- **Industrial applications and energy storage**: Hydrogen is also used in various industrial applications, such as ammonia production and oil refining. It is also used as an energy storage medium, helping

to balance supply and demand on electricity grids.

5.5 Challenges and prospects

- **Economic and technical problems associated with large-scale electrolysis**: Although electrolysis is promising, challenges remain, particularly in terms of production costs and the availability of materials. Research must focus on optimising processes and reducing costs.
- **Environmental impact and strategies for a sustainable process**: The environmental impact of hydrogen production needs to be assessed. Strategies include using renewable energy to power the electrolysers, thereby minimising the carbon footprint of the process.
- **Hydrogen's role in the global energy transition**:Hydrogen can play a key role in the transition to a low-carbon economy, particularly by decarbonising sectors that are difficult to electrify, such as heavy industry and transport.

This chapter has highlighted the importance of water electrolysis for the production of hydrogen, an essential sustainable energy carrier in the fight against climate change. By examining hydrogen technologies, catalysts and applications, we have highlighted the opportunities and challenges associated with this technology. Continued research into catalysts and electrolysis processes will be crucial to realising hydrogen's potential in the global energy transition.

CHAPTER 6

CHALLENGES AND PROSPECTS

This chapter addresses the challenges and opportunities associated with the use of metal complexes in sustainable applications. While these materials play a crucial role in green chemistry and emerging technologies, it is essential to understand current limitations, explore potential innovations, and consider the societal impact of these developments.

6.1 Current challenges : Limitations of metal complexes in sustainable applications

The use of metal complexes presents a number of challenges:

- **Material costs**: Metal complexes, particularly those containing precious metals, can be expensive to produce and isolate. This limits their accessibility for large-scale applications, making economic sustainability difficult to achieve.
- **Toxicity and environmental impact**: Some metal complexes can present risks to human health and the environment. The potential

toxicity of these materials raises concerns, requiring in-depth research to develop less harmful alternatives.

- **Recyclability and sustainability**: The issue of recycling metal complexes is crucial. The lack of effective methods for recovering and reusing these materials in industrial processes limits their long-term durability.

6.2 Innovations and future research: Potential directions for research

A number of promising avenues of research can be explored to overcome these challenges:

- **Development of alternative catalysts**: Research into less expensive and more durable metal complexes, such as those based on nickel or iron, could offer viable solutions. Exploring new ligands and architectures could also improve catalytic efficiency.
- **Recycling technologies**: Investing in efficient recycling technologies for metal complexes will reduce waste and increase the sustainability of industrial applications.
- **Biomimetic approaches**: Taking inspiration from biological systems to design innovative metal complexes could offer sustainable

solutions. For example, natural enzymes have remarkable catalytic characteristics that could be reproduced artificially.

6.3 Societal impact: the importance of social acceptability and sustainability policies

The success of metal complexes in sustainable applications depends not only on technological advances, but also on their social acceptability:

- **Awareness-raising and education**: It is crucial to raise public awareness of the benefits of sustainable materials and the importance of adopting them. Educational programmes could facilitate social acceptance and encourage wider adoption of the technologies.
- **Sustainability policies**: Governments and institutions must establish policies favourable to research and innovation in the field of metal complexes. This includes financial incentives for companies that invest in sustainable solutions and regulations to encourage responsible practices.
- **Interdisciplinary collaboration**: Collaboration between scientists, industry, decision-makers and civil society is essential to developing accepted sustainable solutions. Interdisciplinary initiatives

can foster innovation and ensure that technological developments meet societal needs.

This chapter has explored the challenges and opportunities associated with metal complexes in sustainable applications. Although there are limitations Current technologies, promising innovations and future directions can turn these challenges into opportunities. Social acceptability and sustainability policies will be essential to ensure that these technological advances benefit society while respecting the environment.

CONCLUSION

In this book, we have explored sustainable chemistry and the crucial role of inorganic materials in the transition to a more environmentally friendly economy. Here is a summary of the key points covered:

- **Definition and principles of sustainable chemistry**: We defined sustainable chemistry as a discipline that integrates environmental, economic and social considerations, and discussed the twelve principles of green chemistry that guide this approach.
- **The evolution of sustainable materials**: We have traced the history of sustainable materials, highlighting emblematic examples such as bioplastics and bio-based composites, as well as advances in materials chemistry.
- **Metal complexes and catalysis**: We examined the essential role of metal complexes in sustainable catalysis, exploring their applications in hydrogen production, CO2 conversion and other chemical processes.
- **Electrolysis technologies**: Water electrolysis has been identified as a promising method for hydrogen production, with a focus on catalysts and different electrolysis technologies and their industrial

applications.

- **Challenges and future prospects**: We have identified the challenges associated with metal complexes, including costs and toxicity, while highlighting the following the innovations and future research needed to overcome these obstacles.
- **Societal impact**: Finally, the importance of social acceptability and sustainability policies was discussed as a key element in ensuring the success of sustainable technologies.

Future prospects for sustainable chemistry and inorganic materials

Looking to the future, there are a number of prospects for sustainable chemistry and inorganic materials:

- **Continued innovation**: Sustainable chemistry research will continue to evolve, with innovations in catalysts, bio-based materials and energy technologies. The exploration of new, less expensive and more sustainable metal complexes will be crucial.
- **Integration of technologies**: Synergies between different

technologies, such as hydrogen and electrolysis, and their integration into overall energy systems will play a key role in the transition to sustainable solutions.

- **Interdisciplinary collaboration**: The future of sustainable chemistry will also depend on collaboration between scientists, industry and decision-makers. Working together to develop sustainable and effective solutions to societal challenges will be fundamental.
- **Awareness-raising and education**: Finally, public education and awareness of the importance of sustainable chemistry and inorganic materials are essential to encourage wider adoption of these technologies.

In conclusion, sustainable chemistry and inorganic materials offer immense potential for creating a more environmentally friendly future. The challenges are many, but so are the opportunities for innovation and progress. Continued research and collaboration will be key to turning these prospects into reality.

"ILLUSTRATIVE CHEMICAL REACTIONS

Numéro	- Réaction Chimique	-Description	-Application
1	$NaOH+HCl \rightarrow NaCl+H2O$	Réaction de neutralisation entre un acide et une base.	Illustre le principe de minimisation des déchets en chimie verte.
2	$nCO+(2n+1)H_2 \rightarrow CnH_{2n+2}+nH2On$	Réaction de Fischer-Tropsch, où le monoxyde de carbone et l'hydrogène sont convertis en hydrocarbures.	Utilisation de complexes métalliques pour produire des carburants.
3	$RCOOR'+H2O \rightarrow RCOOH+R'OH$	Hydrolyse des esters, qui peut être utilisée dans la production de bioplastiques.	Création de matériaux biosourcés à partir de ressources renouvelables.

Numéro	- Réaction Chimique	-Description	-Application
4	$CO2+3H2 \rightarrow CH3OH+H2O$	Réduction du dioxyde de carbone en méthanol.	Conversion de CO_2 en produits chimiques utiles, favorisant la durabilité.
5	$2H_2O \rightarrow 2H_2+O_2$	Électrolyse de l'eau, décomposition de l'eau en hydrogène et oxygène.	Production d'hydrogène comme vecteur énergétique durable.
6	$M\text{-}L \rightarrow M+L$	Décomposition d'un complexe métallique, où M est un métal et L est un ligand.	Évaluation de la durabilité des complexes métalliques dans des processus chimiques.

Détails supplémentaires :

- **M** : Métal de transition, souvent utilisé dans des complexes catalytiques.

- ***L**: Ligand, which can influence the catalytic properties and toxicity of the complexes.*

GLOSSARY

- **Biomass**: Organic matter from biological sources, used to produce materials, energy or chemicals.
- **Bioplastics**: Plastics made from renewable resources, such as starch or cellulose, designed to reduce dependence on traditional oil-based plastics.
- **Biobased**: Products or materials derived from renewable resources, such as plants or organic waste.
- **Catalyst**: Substance that increases the speed of a chemical reaction without being consumed in the process.
- **Sustainable chemistry**: Chemical discipline aimed at reducing the environmental impact of chemical products.
- **Green chemistry**: a set of principles and practices aimed at designing chemical products and processes that are more respectful of the environment.
- **Composite**: Material made up of two or more separate components which, when combined, have superior properties.
- **Circular economy**: Economic model that aims to extend the life of products and minimise waste, by encouraging reuse and recycling.
- **Electrolysis**: The process of chemically decomposing a substance by

passing an electric current through it.

- **PEM (Proton Exchange Membrane)**: Proton exchange membrane used in certain electrolysis technologies to improve efficiency.
- **Photovoltaic**: Technology that converts sunlight into electricity using solar cells.
- **Biomimetics**: An approach that draws on biological models to solve human problems.

REFERENCES

1. **Anastas, P. T., & Warner, J. C.** (1998). Green Chemistry: Theory and Practice. Oxford University Press.

2. **Auras, R., et al** (2004). An Overview of Polylactides as Packaging Materials. Journal of Biobased Materials and Bioenergy, 2(1), 1-9.

3. **Benignus, V. A., & Pottenger, L. H.** (2012). Sustainable Development and the Chemistry of Green Chemistry. Nature Chemistry, 4(1), 1-2.

4. **Beller, M., & Bolm, C.** (2004). Transition Metal Catalysis in Organic Synthesis. Wiley-VCH.

5. **Crabtree, R. H.** (2014). The Role of Homogeneous Catalysis in Sustainable Chemistry. Nature Chemistry, 6(2), 115-120.

6. **Clark, J. H., & Macquarrie, J. R.** (2002). Handbook of Green Chemistry and Technology. Blackwell Publishing.

7. **Dicks, A. L., & McLaren, R.** (2010). Chemical and Sustainable Development. Chemistry World, 7(6), 30-34.

8. **Ghadi, R. R., & Goyal, A.** (2017). Sustainable Chemistry and Engineering: A Green Perspective. Journal of Cleaner Production, 141, 123-132.

9. **Giddey, S., et al** (2012). Hydrogen Production from Water Electrolysis: A Review of the Technologies. International Journal of Hydrogen Energy, 37(9), 6748-6754.

10. Green, M. A., Emery, K., Hishikawa, Y., Warta, W., & Zou, J. (2010). Solar Cell Efficiency Tables (Version 35). Progress in Photovoltaics: Research and Applications, 18(1), 1-12.

11. **Grätzel, M.** (2001). Photoelectrochemical Cells. Nature, 414(6861), 338.344.

12. **Kainuma, Y., & Yamada, S.** (2014). Advancements in Bioplastics: A Global Perspective. Journal of Polymers and the Environment, 22(1), 1-12.

13. **Liu, Y., et al.** (2017). Recent Advances in Catalysis for CO2 Conversion. Green Chemistry, 19(22), 5106-5131.

14. **Liu, Y., & Wang, H.** (2019). Recent Advances in Organometallic Catalysts for Solar Energy Conversion. Energy & Environmental Science, 12(2), 463-476.

15. **McDonough, W., & Braungart, M.** (2002). Cradle to Cradle: Remaking the Way We Make Things. North Point Press.

16. **Momo, S. A., & Smith, A.** (2019). Sustainable Materials: A

Review of Bioplastics and Composites. Journal of Cleaner Production, 211, 562-577.

17. **Mohanty, A. K., Misra, M., & Drzal, L. T.** (2002). Sustainable Bio- Composites from Renewable Resources: A Review of the Progress. Composites Science and Technology, 62(12), 1467-1482.

18. **Mohanty, A. K., & Misra, M.** (2016). Sustainable Biodegradable Composites from Renewable Resources: A Review. Journal of Polymers and the Environment, 24(4), 1-18.

19. **Pollard, S. J., & Macleod, M.** (2007). Sustainable Development and the Role of Chemistry. Chemical Society Reviews, 36(6), 1045-1057.

20. **Ragaert, K., Delva, L., & Van Geem, K.** (2017). Mechanical and Chemical Recycling of Solid Plastic Waste. Waste Management, 68, 120- 133.

21. **Su, Y., et al.** (2019). Innovative Strategies for the Development of Sustainable Catalysts. Chemical Society Reviews, 48(6), 1313-1332.

22. **Thakur, V. K., & Thakur, M. K.** (2014). Biodegradable Polymers and Their Applications. Composites Part A: Applied Science and Manufacturing, 67, 1-12.

23. **Turner, J. A.** (2004). A Realizable Renewable Energy Future. Science, 305(5686), 972-974.

24. **United Nations.** (2015). Transforming Our World: The 2030 Agenda for Sustainable Development.

25. **United Nations Environment Programme.** (2021). Policies for Sustainable Development.

26. **Wang, Y., et al.** (2019). Advances in Catalysts for Hydrogen Evolution Reaction in Electrolyzers. Chemical Society Reviews, 48(9), 2317-2339.

27. **Liu, Y., & Wang, H.** (2019). Recent Advances in Organometallic Catalysts for Solar Energy Conversion. Energy & Environmental Science, 12(2), 463-476.

MIX
Papier aus verantwortungsvollen Quellen
Paper from responsible sources
FSC® C105338

Printed by Books on Demand GmbH, Norderstedt / Germany